Climate Change and You

Facts and Myths

Climate Change and You

Facts and Myths

Roger McCallum

Copyright 2024, Roger McCallum

Print Book: ISBN # 978-1-0688321-2-3
E-Book: ISBN # 978-1-0688321-1-6

Printed by lulu.com

Table of Contents

Introduction

The weather is not what it used to be. The climate is changing, and we are going to have to make the most of it. This article explores the impact you have on our climate, and the idea that we can control the climate by adopting green energy solutions.

To begin with, 80% of the world's energy is supplied by fossil fuel. Fossil fuel consumption has increased every year since the 60's. There is the impression that green energy is on the brink of taking over. This is a misconception. After 20 years of promotion, green power has not reduced the rate of increase in global consumption of fossil fuels.

Our society is completely dependent on fossil fuel and that energy source is running out. We are not doing a very good job of finding a replacement. In spite of this major problem, international attention has not been devoted to developing a solution. Our attention has been focused on the symptom.

We are focusing on reducing CO_2 emissions. The focus should be on finding a replacement for fossil fuel.

Many people will state with certainty that CO_2 (carbon dioxide) causes climate change and we have to think of future generations. If we run out of the energy source that supports our civilization before we find a viable alternate, climate change will be the least of our descendants' problems. Finding an energy source to reduce fossil fuel consumption will, as a side effect, lower CO_2 emissions.

Replacing fossil fuel is a collective problem to which there is a solution. The solution does not require massive new public spending, or new tax. It simply requires a re-orientation of policy.

The idea that CO_2 is going to destroy the world and is the root of all evil is, in my opinion, an exaggeration. I see CO_2 and its association with climate change designed more to separate you from your money than as a solution to a global problem. We are going down a road paved with good intentions. People who have an interest in your money just need to give the good intentions a little nudge from time to time.

The sections that follow will explain my point of view supported by facts. The data presented in this article is available to anyone. You are welcome to check my numbers. Sample calculations and any assumptions appear in the appendices.

1) CO2 by the numbers. Where does it come from and where does it go?

Almost all fauna (animals) inhale oxygen and exhale carbon dioxide[1]. CO_2 is released when plant matter rots, which is essentially worms and bacteria breathing. CO_2 is produced when wood, plant mater, or fossil fuel combines with oxygen and burns. The largest source and sink (place where CO_2 goes) is geological. CO_2 finds its way into sedimentary rocks like limestone, and is eventually released by volcanoes. It is also released when we heat limestone to make cement.

CO_2 does not stay in the atmosphere. All flora (plants) from trees to algae in water absorb CO_2. Using energy from the sun, plants use the carbon in carbon dioxide to make cellulose and other plant matter. The process called photosynthesis. Oxygen removed from carbon dioxide is released into the air by plants. It is because of this cycle that we have oxygen to breathe. Carbon dioxide is an essential part of that cycle.

The amount of CO_2 in the earth's atmosphere is between 350 and 400 parts per million (ppm), which can be expressed as .04%. CO_2 is a minor constituent of the atmosphere, which is about 21% oxygen and 78% nitrogen. The atmosphere also contains about 0.25% water vapour. CO_2 absorbs sunlight which warms the atmosphere. This is why CO_2 is called a greenhouse gas. Water vapour also absorbs energy from the

---

1 Carbon dioxide is a colourless, odourless gas containing one atom of carbon (C) and two atoms of oxygen (O) the chemical formula is CO_2

sun and is a greenhouse gas.

Even at .04%, there is a lot of CO_2 in the atmosphere. Approximately 2,000,000 million metric tons[2].

Table 1-1 shows how this number compares with annual CO_2 emissions from fossil fuel by fuel source and emitting country. CO_2 emissions are calculated from published energy consumption figures and factors calculated by the EIA, a US government agency, telling us the amount of CO_2 generated by fuel type based on energy delivered. A sample calculation appears in appendix 1.

Table 1-1: Calculated CO_2 generated in 2022 from hydrocarbons showing the fuel type.

Region	Fossil Fuel type	2022 exajoules	million metric tons CO_2	Fossil total Million metric tons CO_2	Percent of world's fossil fuel CO_2
world	coal	161.47	14,680		
	natural gas	141.86	7,114		
	oil	190.69	13,571	35,365	
			0		
China	coal	88.41	8,038		
	natural gas	13.53	679		
	oil	28.16	2,004	10,720	30.3
			0		
India	coal	20.09	1,826		
	natural gas	2.09	105		
	oil	10.05	715	2,647	7.5
			0		
Canada	coal	0.39	35		
	natural gas	4.38	220		
	oil	4.27	304	559	1.6

---

2 You may prefer 2×10^{12} metric tons. I express these unimaginably large numbers as millions of metric tons so the reader can easily make comparisons.

Fossil fuels are not the only source of CO_2. Every time you inhale, you breathe air containing 0.04% CO_2. You exhale air containing 5% CO_2. Using standard lung capacity and a number of assumptions, every human on earth exhales about 2.8 kg of CO_2 every 24 hours. Table 1-2 shows calculated annual CO_2 generation for various populations. Calculations and assumptions appear in appendix 2.

Table 1-2: CO_2 produced by people breathing for a year

region	Population - Millions	millions of metric tons CO_2 per year
Canada	39	39
world	8,000	8,094
China	1,430	1,447
India	1,430	1,447

Other natural phenomena generate CO_2. In Canada during 2023, massive fires burned 14 million hectares of forest. CO_2 generated during the year from burning forests was approximately 4,500 million metric tons.

There is more CO_2 stored in rocks than is found in fossil fuel. CO_2 is released by volcanic activity. The US geological survey report that in a typical year, from 180 to 440 million metric tons of CO_2 are emitted by volcanoes.

It is widely reported that CO_2 is responsible for climate change. There may not be such a clear one-to-one relationship, which is the subject of the next section.

Botanical research has shown that as CO_2 levels rise, plants become healthier and larger. This same research suggests that ice-age levels of CO_2 may have made conditions difficult for plant life to survive. Increasing CO_2 levels will help agriculture and plant growth. Increased global foliage will absorb more CO_2. Eventually an equilibrium will be reached.

CO_2 levels in the atmosphere have increased over the past

100 years. Who is to say the levels of 100 years ago were normal? During the Mississippian and Pennsylvanian period (359 to 259 million years ago) when the coal fields were developing from forests that grew so fast they did not have time to rot, CO_2 levels were thought to have been at least .4%, 10 times higher than today's levels. CO_2 levels dropped all by themselves to .03% or .04% ppm as the Permian ice age started. This lasted from 350 to 225 million years ago when CO_2 levels rose by themselves without human intervention to about .2%, 5 times today's levels. This was the average during the Jurassic period when dinosaurs ruled the earth. CO_2 levels then fell gradually to about 0.03% as the last (Tertiary) ice age began.

We can set these facts aside for the moment and consider CO_2 production numbers, and where CO_2 is produced. This calls into question the story that individual contributions of Canadians can save the planet. The population of India, by simply living and breathing every year, generates almost three times the amount of CO_2 Canadian homes, businesses and industry emitted by burning fossil fuel during 2022. In addition, China produces 30% of the world's fossil fuel generated CO_2 compared to Canada's 1.6%. China does not participate in global conferences or commitments on global warming.[3]

How can your contribution and sacrifice control the climate of the world? Canadian forest fires in 2023 produced about ten times the amount of CO_2 emitted by all fossil fuel consumed in Canada during 2022.

In spite of clear data to the contrary, we have all heard that we are responsible. Living responsibly is a buzzword, which translates to "I am better than you, or I would not have to tell you to live responsibly."

This is the same message delivered by old-testament proph-

---

3 Chinese people are not villains. Heavy industry generates a lot of CO_2, but gives the people living in the industrialized nation an un-paralleled life style. China is following a well-trodden path to prosperity, one already taken by most western nations.

ets who promised their followers victory in war if everyone said the appropriate prayers and lived a pious lifestyle. When they lost in battle, the prophets told their followers that they were not righteous enough to deserve God's protection. Looking back at the events, historians can see that God and prayers had nothing to do with the events. Had the biblical population properly assessed the situation, they would have seen that their armies were out-manned and out-gunned. Rather than look to God for a miracle, they should have avoided the fight. Praying, and feeling guilty about their failures did not help.

CO_2 is used to assess personal guilt. Building an energy infrastructure that replaces fossil fuel and at the same time reduces CO_2 emissions is a government project or a big business project, or the work of massive utility companies. There is nothing to feel guilty about. If CO_2 rather than the real problem becomes the focus, people can be made to feel guilty. Buying an electric luxury car and having a detached home so you can charge the car overnight is not extravagant. It is responsible. Those who cannot afford an electric car will be more willing to pay carbon tax as a form of penance. People feel better about paying the inflated prices in the grocery store caused by high fuel prices, carbon tax and carbon offset fees because it is for the common good. Oil companies are under no pressure to lower prices because like cigarettes, fuel has become a dirty luxury. It would not be responsible for the government to intervene. Industries such as steel and cement that cannot produce their product without generating large amounts of CO_2 will have to buy carbon credits. Because of our guilt, we will accept the extra cost on these essential products that will inevitably be passed on to us. Every extra nickel a tin can costs, every dollar of tax increase to repair our sidewalks will be concentrated into the pockets of cap-and-trade brokers, a new breed of billionaire.

This is because our attention has been nudged away from the real problem. We are running out of a resource that is essential to our society. The only part of the problem we hear about is CO_2.

2) What controls the weather

The modern concern with CO_2 levels in the atmosphere is based on climate models, and in particular the IPCC (Inter Governmental panel on Climate Change) written and updated from time to time by a United Nations agency. The document is a scholarly work containing more than 360 pages. You can download a copy of the report for no cost.

The report is a collection of articles written by unimpeachable experts in their fields. The articles make predictions describing possible effects of CO_2. This was what the authors were asked to do. The articles forecast the effect rising CO_2 levels may have on vegetation, precipitation and weather. The original report contained the word "may" 320 times.

The word "may" occurs frequently because climate is a complex subject that is not fully understood. This is a fact stated in the IPCC report. Climate is the result of interactions between many variables, not just CO_2 levels. IPCC researchers were asked to comment only on the effects of CO_2.

Weather and climate are complex subjects, yet CO_2 is being held up as the single source of climate change. There is no scientific consensus as to cause and effect. Yes, the climate is changing, and yes, CO_2 levels are higher than they were in preindustrial times. What is not clear, and what no scientist has proven is that one thing has anything to do with the other. Models, not our understanding of the atmospheric interactions provide grave warnings about the effect of increasing CO_2 levels. Models are only as good as our understanding. We do not understand the interaction well enough to be able to predict the weather with certainty. It is hard to understand how we have concluded that a gas that makes up an insignificant amount of the earth's atmosphere can, as a single variable, control climate. If the relationship were that simple, you would have an app on your phone that would tell you with certainty what the weather was going to be on any date, in the location you selected.

Weather forecasting is not that easy. The long-term forecast can tell us, for example, that there will be a lot of snow next winter, or that the upcoming summer will be hot and dry. If this prediction is dead wrong, we do not hold the weatherman responsible. Not much is said after the fact because we do not expect anyone to be able to predict even general weather patterns six months into the future. Any wedding planner will tell you short-range forecasts cannot predict the weather one week into the future with certainty.

In spite of what we all know, governments are basing policy and imposing new taxation on a forecast that claims to be able to predict the weather based on the effects of a gas that makes up only 0.04% of the atmosphere.

Taking a different approach, water vapour is a greenhouse gas. Water vapour in the concentrations found in the atmosphere has the same heat absorbing capacity as atmospheric CO_2 levels. We all know that as temperature rises, water evaporation increases, which will increase the amount of water vapour in the atmosphere.

Common sense suggests that as the planet heats, water vapour concentration will rise. This will allow the atmosphere to absorb more heat. This describes an out-of-control feedback loop that would result in an exponentially increasing global temperature. In spite of what we think we know, this out-of-control heating does not take place.

Global temperature finds an equilibrium. There are many factors that govern the equilibrium.

One of these factors is that when the earth is facing away from the sun (the night) heat is radiated out into space. The rate at which heat is radiated from any object is controlled by complex factors and constants, but radiative heat transfer increases dramatically as temperature increases. The effect of water vapour on atmospheric heating is complex and one that is not dealt with at length in the IPCC report because it is not well understood.

There is an equilibrium that controls CO_2 levels. During geological history of the earth, the CO_2 levels have risen and fallen all by themselves as the equilibrium shifts. The factors that control this equilibrium are not fully understood.

There is more to climate change than the CO_2 level.

The climate is changing. The climate has been changing throughout history. The climate changed in prehistoric times. It is accepted that early humans crossed what is now the Bering Strait between Russia and Alaska beginning 30,000 years ago and ending 16,000 years ago. This was possible because sea levels were lower. All that water was tied up in ice that covered much of Europe and North America. The Bering Strait is now 60 feet deep. Sixty feet in 16,000 years is a lot of sea level rise. The average increase in water level is over 1 mm per year. We know that Australia was connected to Tasmania, Indonesia, New Guinea and Thailand so that early humans and fauna could walk to these places. They were later isolated by rising sea levels.

Polar ice caps are still melting. It should come as no surprise to be told that sea levels are rising. This does not mean the sky is falling.

Cave paintings in North Africa show animals living in what is now the impenetrable Sahara Desert. There are clear signs of water erosion around the Sphinx and the barren hills of Egypt that could only have been caused by significant amounts of rain.

Major climate changes took place before the pyramids had been built. It should alarm no one to be told today that our climate is changing.

We have to be prepared to face a changing climate. We cannot control the climate, but if we are to survive, we must be able to adapt.

In spite of this, we are told that if everyone does their little bit to reduce greenhouse gas generation, we can control the climate. We do not need to do anything other than be good. If we are good enough, the sea will stop rising and nothing will ever change again. The prophets making these pronouncements do not specify if the climate will remain as it is today, go back to what it was in the 1980's, 200 years ago, or 500 years ago. These things are not the same, and like the good old days, they probably did not seem so good at the time.

If you listen to me and I am wrong, will you be destroying the planet? That is unlikely. The planet was struck by a meteor the size of Manhattan 65 million years ago. This released more energy than would have been released if all nuclear weapons ever made had been detonated at once. It is estimated the impact resulted in 700 mile-per-hour winds with air temperatures reaching 400 degrees sweeping around the entire globe.

The planet was not destroyed.

The dinosaurs were not so fortunate.

We cannot destroy the planet. It is possible that a changing climate will destroy life as we know it, but not the planet. All penance performed to stop climate change is in reality an effort to save our lifestyle, and possibly our civilization.

Important as that may be, this does not sound as noble as saving the planet.

Our lifestyle and our civilization can only be saved it we find a substitute for burning fossil fuel before there are no viable reserves left. If we run out of fossil fuel before we have a well-developed alternate, our lifestyle and our civilization will surely be destroyed. If we can lower CO_2 emissions at the same time all the better, but it will be the lack of energy, not CO_2 emissions that has the potential to cripple future generations.

What about extreme weather? The history of the world is a history of weather extremes. Extremes are not new events. Hurricane Hazel struck Toronto in 1954. Eighty-one people were killed and 1,900 families were left homeless, some houses were washed into Lake Ontario.

The great depression in the 1930's may have started with a stock market crash, but it was successive years of drought and crop failures across western Canada and central United States that caused real misery for millions of people who had never invested in the stock market. Years of drought that caused the dust bowl forced people to abandon farms and entire communities to disperse.

There were times in the earth's history when there was no ice on either pole. We are heading in that direction. The water has to go somewhere. Sea levels are going to rise, yet we still issue building permits near low, coastal areas. This is a recipe for disaster. It will be called a climate disaster, but it is really a planning disaster.

Hurricane Katrina in 2005 flooded sections of New Orleans. Some of those developments were built below sea level. Calling this event a climate change disaster had not come into vogue at that time, but today it would be another example of climate change for which you are guilty. Building permits were issued to build houses on land that was below sea level in a coastal city. The loss of property and life was a planning disaster. The old part of the city, the French Quarter, was not damaged. Even in the 18th century, they knew enough to build on high ground and without basements.

During 2022 we saw images of extensive flooding in the Fraser Valley in British Columbia. This was reported as being a climate event, and proof that we needed carbon tax.

What was not reported was that major floods had occurred in the Fraser Valley as far back as 1894 and periodically ever since, including 1948, 1972 and 2007. Sumas Prairie was once the site of Sumas Lake. The lake was drained by a developer in 1929 as part of a massive project that included diverting the Chilliwack River. Much of the Fraser Valley was swamp and marsh until this draining.

Flooding has occurred in the region since it was settled. Serious as it was, it cannot be called a climate disaster. Rather than using this as a hand-wringing platform calling for sanctions against all Canadians, a more appropriate government response would have been to invest in a floodway to divert water the next time there is a flood.

Winnipeg had a major flooding problem going back to the days of early settlement. Over the years, they built floodways. Many houses in flood plains were condemned and demolished. New construction building permits in low areas were not issued. They solved the problem before anyone had heard of climate change.

We have to prepare ourselves for extreme weather events if we are to survive. Contributing to cap-and-trade brokers will not help. The notion that climate change is your fault and you can fix it belongs in the Old Testament. There has been a suggestion that Neanderthal man went extinct because they could not adapt to the changing climate toward the end of the European glaciation during the last ice age. Let's hope we can do better.

We cannot control the climate, or bring back the good old days, but we can mitigate flood risk. Clear-cut deforestation still takes place in Canada. Our government has chastised the Brazilians, telling them to do more to protect the rain forest. Canadian forests are being decimated at an alarming rate. Canadian forests do not grow back as quickly as those in Brazil. Forests prevent erosion and hold water. Wetlands absorb water when it rains. Deforestation often results in mud slides and flooding. Every time a wetland area is drained, we intensify flooding.

A good quality affordable house or high-rise can be built using steel and concrete. We are villainizing the steel and concrete industries because it is not possible to produce their products without generating CO_2. We still allow clear cutting to provide building materials. Trees remove CO_2 from the air. Is it better, or simply more profitable to make a company pay to produce CO_2 than to save forests that take CO_2 out of the atmosphere?

There is a strong indication that if the warming trend continues, some highly-populated parts of earth will soon be uninhabitable. This is made out to be a man-made catastrophe. It has happened before. Thriving prehistoric civilizations in the Gobi desert and in the Sahara were either wiped out or they moved. There is no indication this early climate change was the result of Neolithic industrialization.

When regions of the world become uninhabitable due to climate change, I hope we will not be looking at internet feeds and wringing our hands and assessing blame on the dirty rich nations for not living responsibly. World leaders have to live up to what they are called. They will have to lead.

Our society has to develop a plan to move people out of the areas where the temperatures are inevitably going to exceed 40C. We have room, and it will be easier to help everyone move than to pay for wars. Fortunately, the Dubai climate conference in 2023 included some strong provisions for assisting people affected by climate change. Industrialized, wealthy nations are contributing to a common pot. Although I maintain that industrialized nations are not as guilty as we have been led to believe, the resolution to have the rich helping the poor could save us from generations of devastating world conflicts.

I believe that the real tragedy of the dust bowl of the 1930's was the lack of government support for millions of farmers and small business owners devastated by the extreme weather of the times. Common people were abandoned. There was no money to help. A few years later the same governments found enough money to finance a world war.

If we can take the initiative to help people in need caused by extreme weather, we could prove that we are can rise above the level of Neanderthals.

3) Why do we use fossil fuel

This is a look at why we use oil gas and coal. You will be able to see the amount we use, and better understand why the use of fossil fuel alternatives have not achieved widespread acceptance.

Chart 3-1 shows how dependent we are on fossil fuel.

Chart 3-1: Global energy consumption by source. Data source: Statistical Review of World Energy data

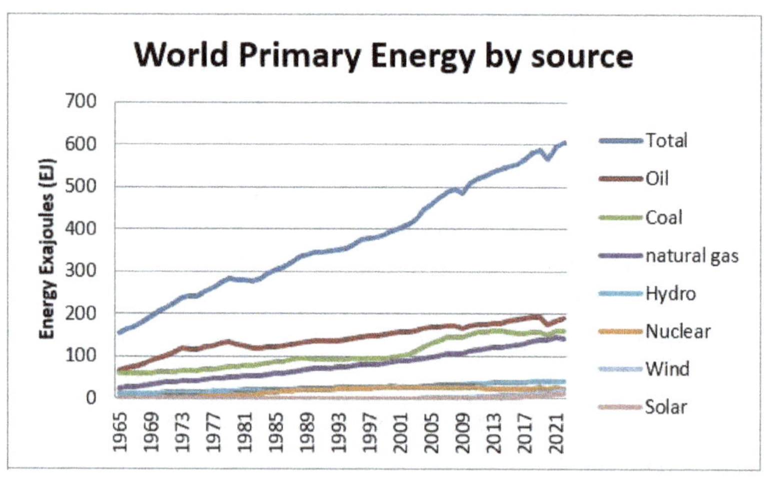

Global energy consumption has grown steadily for the period charted, from 1965 to 2022. The use of fossil fuel has contributed more than any other energy source to that increase.

Renewable energy does not play a significant role in global energy consumption. Fossil fuel provided 81 percent of the world's energy in 2022, while combined wind and solar accounted for only 5.3%.

If these numbers are divided by population, the growth is not as pronounced. Energy consumption goes up, but some of the growth in energy consumed can be explained by the increase in world population.

Chart 3-2: Global per capita energy consumption

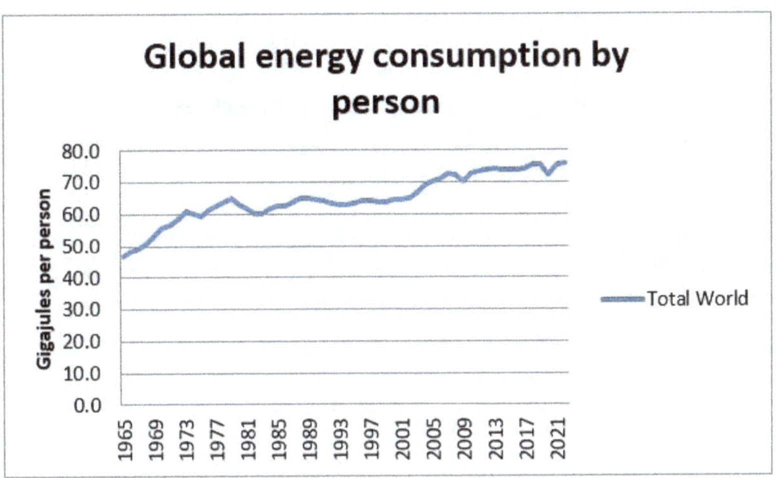

Developed countries use more energy per person than less developed countries, and there is a direct relationship between energy consumed in a country, quality of life and life expectancy. Energy is required to provide fresh water, treat sewage build roads and hospitals. While some of the increase in global energy consumption has been driven by population growth, a part of the increase has been driven by an increase in the standard of living in less developed countries through energy consumption.

Chart 3-3 shows the difference between some developed and developing nations in terms of energy consumption. Converting to more familiar unit of energy, 400 Gigajoules (GJ) equals over 100,000 kWh. This is a massive number, but it is all energy consumed by a country for a year, divided by the number of people. It includes not only personal energy consumption such as electricity and gasoline, but also energy to run steel mills, operate freight trains and all other energy consumed in a country divided by population.

Chart 3-3: Per capita energy consumption of selected countries.

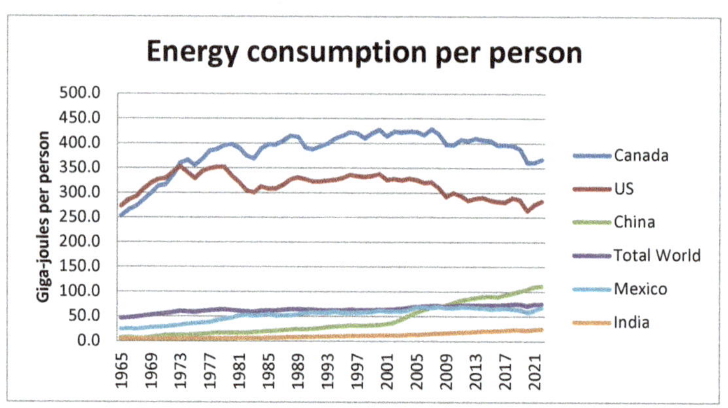

Chart 3-3 shows the disparity in energy consumption around the globe. Some of the disparity in developed counties is due to industrialized countries producing manufactured goods for export.

How did we get into this position of total dependence on high density forms of energy?

Looking at graph 3-1 and extrapolating back, it is clear that not long ago we did not need any of the energy forms we have been discussing. Major civilizations existed before we started using petroleum products. Coal only became important once the industrial revolution had started in the 17th century. Oil was not extracted or refined until the middle of the 19th century. Oil was first extracted to produce kerosene, a low-cost substitute for whale oil used in lamps.

No energy that is described in Chart 3-1 was used in the classical world because the energy source of the ancients was slavery.

Slaves rowed boats, performed all the tasks now performed by machines in mines, on construction projects and in farming. Rome was constantly at war, and the main purpose of those wars was to enslave the people they conquered to provide energy for their imperial requirements. Slavery was not confined to the great cities of Rome, Carthage and Constantinople. Slavery was part of rural and nomadic life during the Roman period. Slavery is described, and clearly endorsed as a way of life in the bible. Hannibal attacked Rome by driving elephants through the alps. Roman soldiers he defeated were sold as slaves to help finance his armies. When Hannibal's army was eventually defeated at the gates of Rome, the survivors were captured and sold by the Romans. No one got to go home when it was over.

After the fall of the Roman empire, slavery continued in the Ottoman empire and other sophisticated cultures. European societies were not as sophisticated, and did not require all the services that Roman slaves provided. None the less, Europeans used slaves. Slaves are mentioned in Beowulf, the oldest legend from a European culture. It was recorded in writing during the 6th century. During 1575, M. Cervantes, the author of Don Quixote was captured by pirates during a voyage near Barcelona and sold as a slave in Algiers. He remained in captivity for five years. A more familiar institution than slavery in mediaeval Europe was serfdom. This institution, although not as severe as slavery, forced people to work for land owners who had absolute authority over their lives.

Slavery was a way of life since the beginning of recorded

history. The American Civil War was fought over slavery, and is often seen as the event that ended slavery. The Civil War was an important event, however the driving force that enabled society to put an end to slavery was economic. Slavery could end because there was a viable alternative in the form of fossil-fuel powered machinery. Not only was it viable, but it was less expensive. A barrel of crude oil (defined as 42 US gallons) contains 5.6 million BTU which is equivalent to 5.91 Gigajoules. In order to produce the energy contained in a single barrel of oil, a person in good physical condition would have to work 8 hours a day for 5 years. Thousands of people pulling on ropes could be reduced to hundreds with a machine powered by a steam engine.

Our society did not develop better morality; the slave owner simply could not compete with a steam engine.

We have to find a substitute for fossil fuel. It is a non-renewable resource, and before long, we will run out. If we do not want to condemn our descendants to a life of slavery, an alternative is required. Chart 3-1 shows that solar and wind power are not coming close to replacing fossil fuel. In spite of incentives, subsidies and public encouragement, these 'green' energy sources are not even keeping up with increased energy needs of the world.

The advantages of fossil fuel must be examined to understand how it can be replaced. Wishing will not help. It will take more than an international resolution and punitive measures to significantly reduce an energy source that accounts for 80 % of energy consumed on earth, and essentially supports our entire civilization.

Why can we not stop using fossil fuel?

In spite of the clear fact our supply of fossil fuel is limited, we keep using more every year. There are several reasons.

High-energy density

A lot of chemical energy is packed into gasoline and diesel fuel. Ethanol added to gasoline contains 30% less energy than an equivalent volume of gasoline. Attempts have been made to use biofuels in the aviation industry, but these fuels do not contain the same amount of energy as jet fuel, so the engines develop less power and the plane has less range.

The most striking example of energy density is found with electric cars. When you fill a 50-litre tank of a gas car you are putting the equivalent of 500 kWh of stored energy in to the car. 50 litres of gasoline weighs 37 kg. The storage container can be fabricated for about $50. The tank is the size of a suitcase. It takes longer to pay for the 500 kWh of energy than it does to put it into your car. If you live in a remote area, you can carry a spare container of gasoline or keep a reserve at home for almost no storage cost.

This is not the case with an electric car, where energy density is a real problem. State-of-the-art battery technology costs over $200 per kWh of storage. The cost of a 250-kWh battery is in the order of $50,000. The battery weighs as much as 700 kg, or 1,500 pounds. For all this money and weight, the battery holds the same amount of energy as a half tank of gas.

Charging times are a big problem with electric cars. In order to achieve rate-of-energy delivery similar to that of a gas car, car you would need a power supply capable of delivering 15,000 kW. This would give you a two-minute fill time, but it represents a technically impractical power delivery rate. Due to better efficiency, a battery charge of 200 kWh should go the same distance as 500 kWh of gasoline. Even with this taken into consideration, you could not put electrical energy into the starship Enterprise as fast as you can put chemical energy stored in gasoline into your car.

Contained energy

An important benefit of fossil fuel is that it contains more energy than it takes to extract and deliver the fuel. Some forms of alternate energy fail in this regard. An example is hydrogen. Hydrogen is all around us in the form of water. Although hydrogen is a gas, there is no free hydrogen in the atmosphere.

Water can be broken down into its component gasses, hydrogen and oxygen. It takes more energy to break water down into its component gasses than you get out of the gasses when they are combined and ignited. If this were not the case you could break down water and burn it in an engine that generated electricity to make more hydrogen and oxygen. This would be a perpetual cycle, or a perpetual motion machine.

Hydrogen is generated from electricity for special purposes where money is not as important as having the gas available for an experiment. Sadly, hydrogen is also generated from electricity by companies that want to add 'green' hydrogen into natural gas to get carbon offset credits. We would be better served by using the electricity directly.

Ninety percent of the world's hydrogen is produced by cracking methane gas. Cracking methane produces hydrogen and carbon dioxide. Hydrogen is a light gas. For every ton of hydrogen produced, 11 ton of CO_2 must be released to the atmosphere. Calling hydrogen a green fuel is misleading. We would be better served if we burned methane (the major component of natural gas) directly. It would produce the same amount of CO_2 as burning hydrogen, and we would not have the losses that are associated with the process of cracking methane.

There may be a glimmer of hope. It may be possible to harness unused wind and solar power to produce hydrogen. This would be a superior storage method than considering battery storage.

A method is being developed to produce hydrogen as a by-product of nuclear power generation.

Contained energy in ethanol

Ethanol takes almost as much energy to produce as it contains as a fuel. This concern was raised during the 1990's. A detailed study found that ethanol production consumed more energy in the form of farm fuel and fertilizers than the ethanol yielded. A lot of work has gone into the farming methods, and more recent data shows that today, ethanol contains 1.3 times the energy needed to produce the fuel. As a comparison, gasoline contains 20 times the energy it takes to extract and refine the product. Oil industry sources note that this has dropped from the early 19th century. At that time, gasoline contained as much as 100 times the energy it took to produce. Contained energy has remained the same. The reduction is caused by increasing energy requirements to extract fuel from dwindling reserves.

How does fossil fuel compare to wind power?

Wind power takes a lot of space, and there is an environmental concern with their installation. Wind farms are often built on flat land close to existing infrastructure. This describes land that is potentially good for agriculture or housing. The small individual units that make up a wind farm are hard to maintain. A wind farm may have hundreds of turbines, all high in the air, outdoors and spread out over miles of property. A fossil power station may have 5 or 6 large turbines, all in an enclosed space and under one roof. The power output can change as demand changes by regulating the amount of fuel. The most important aspect of traditional power generation is that it will produce power when power is needed. Wind turbines are subject to the wind that does not always blow when we want.

Fossil fuels and solar power

Solar power is a good fit for hot, barren areas, however it does not provide power at night, and the energy produced cannot be practically stored. In a northern country like Canada there is not as great a solar radiation flux as there is closer to the equator. Solar power generation suffers from the same problem as wind generation. The technology cannot produce power on demand.   Once again, there is a tendency to install this form of collection device on prime land, rather than find locations that cannot be put to other use.

Solar fields are located outdoors and is made up of hundreds of individual collectors. These are subject to environmental problems including hail, snow, dust and pollen that will reduce output. Solar panels deteriorate over time and cannot be repaired. Solar panels have a limited life and there is no clear path for recycling.

The problem of green energy being available when it is not needed cannot be solved by storing the energy for later use. Storage of electric energy is not practical. Car-size batteries are stretching the limit of battery technology. The energy required to manufacture a lithium ion battery including mining, transportation of raw materials and manufacturing is approximately 240 kWh per kWh of storage. It is difficult to estimate the required capacity to store sufficient energy to supply a power grid. As a reference, the manufacture of a single 100 kWh car battery will consume 24,000 kWh[4]. This is nowhere near the storage capacity that would be required to serve as a buffer between intermittent green power generation and the electric grid. The energy cost to produce such a massive storage device would be astronomical, and would have a matching price tag.

---

4 24,000 kWh, the energy required to produce a single 100 kWh electric car battery is about the amount of energy consumed in a year by a large house in Canada. This would be the equivalent of 2,300 liters of gasoline.

A 100 kWh car battery costs at least \$30,000. It is not difficult to understand why talk about using battery banks to store energy produced by wind or solar rarely get beyond the talk stage.

Fossil fuels do not have this storage problem. Coal can be piled on the ground and remain outdoors for years if necessary. Liquid fuels such as gasoline and diesel can be stored in relatively inexpensive steel storage containers almost indefinitely. At the end of their life cycle, the tanks can be cut up and delivered to a steel mill to be melted, usually generating enough money to pay for the scrapping and transportation.

The only energy source that is a contender to replace fossil fuel is nuclear energy. There is a section devoted to nuclear energy.

4) Poor performance of 'green' energy production

We have to reduce fossil fuel consumption before it runs out and our society hits a brick wall. For the past 20 years, wind power and solar power have been promoted as the solution. Published data makes it clear that 'green power' is not even reducing the rate of fossil fuel consumption, much less leading us toward the elimination of our need for fossil fuel. This is illustrated in Graph 1 compiled from data provided by the Statistical Review of World Energy.

Graph 4-1: Primary energy consumed in Canada by energy source. Units Exajoules.

The data is graph 4-1 is for Canada, however as we have seen from chart 3-1 the graph plotted for world consumption has a similar appearance. In 2022, 64 % of primary energy consumed in Canada came from fossil fuel. Some of this is used directly, for example to heat homes. Some primary energy is used to generate electricity. Wind and solar provided

6.9% of Canada's electric energy in 2022, but this represents only 2.9% of Canada's primary energy consumption.

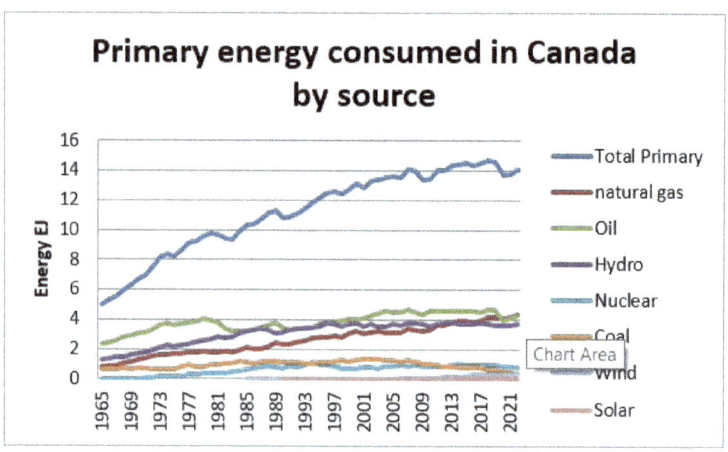

This growth in renewables has come after more than 20 years of subsidies and incentives. A large investment has been made in wind and solar technology, however it does not appear to have been money well spent. In 2022 we benefited from only 27.5% of installed wind capacity.

Table 4-1 lists electrical energy consumed in Canada during 2022 by source.

Table 4-1: Electric energy production and installed capacity in Canada 2022. Data: Statistics Canada, and Statistical Review of World Energy and industry announcements

	Produced 2022		installed capacity 2022	Utilization
	TWh³		GW	%
fossil	117.76		36.2	37.1
nuclear	86.6		14.5	68.2
solar	5.95		4.32	15.7
wind	37.15		15.4	27.5
hydro	394.5		82.2	54.8

Electric energy has to be used when it is generated. When you turn on a light or a stove, a generator connected to the grid has to work a little harder to provide the power you have started to draw. Wind turbines cannot be relied upon to provide power when it is needed. Peaks occur when it is very hot or very cold. Wind velocity is often low during these extremes. Solar should be a good match for hot daytime conditions, but our use of only 15.7% of installed capacity suggests the availability of solar power does not match the demand very often.

In 2022, installed wind generating capacity was greater than nuclear generating capacity. Table 4-1 shows that nuclear plants provided more than two times as much energy as wind. This is because the nuclear plants can deliver power when needed. Graph 4-1 suggests that had the use of nuclear energy increased on the trajectory it had been on in 1990, as much as 30% of our energy could have been nuclear, replacing that much non-renewable fossil fuel.

5) Another look at nuclear energy.

In 2022, only 4 % of the world's primary energy was delivered by nuclear power stations. Nuclear energy supplies 5.5 %

of Canada's primary energy needs, while in the United States it makes up 7.6 %.

Nuclear power plants are similar to a coal or oil generating station. In place of a boiler that burns coal or oil, a controlled nuclear reaction generates the heat. Heat exchangers isolate the water, steam and condensate from the nuclear reactor.

Graph 5-1: Nuclear energy as a percent of primary energy. Data source: Statistical review of world energy

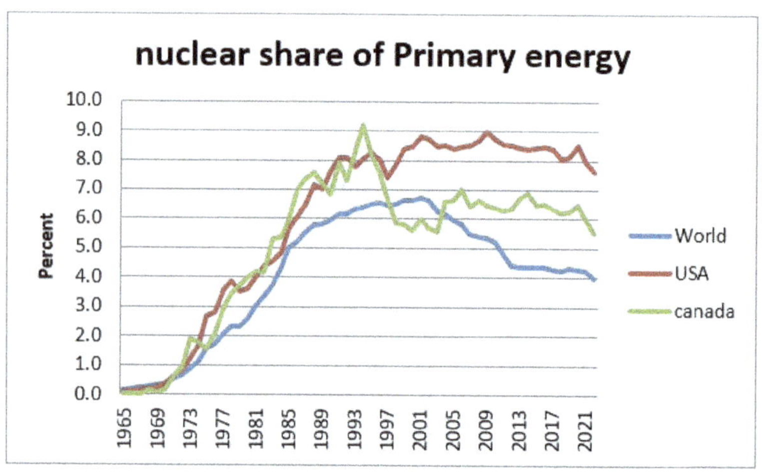

Graph 5-1 shows that in Canada, and around the world, nuclear energy was playing a rapidly increasing role from the early 1970's up until the 1990's. At that point, the rate of increased use of nuclear energy leveled off or, as we see in Canada and most of the world, it dropped.

Looking at the period from 1975 to 1993, the increase in energy supplied by nuclear reactors increased at a rate greater than the overall increase in energy consumption. This can be seen in Table 5-1.

Table 5-1: Increase in nuclear energy use and total energy consumed from 1975 to 1993

	1975	1993			1975	1993	
	Nuclear EJ	Nuclear EJ	Increase (%)		Total Primary EJ	Total Primary EJ	Increase (%)
World	3.8	22.3	490.8		240.8	351.3	45.9
USA	1.9	6.6	253.8		69.6	84.2	21.1
Canada	0.1	1.0	642.1		8.2	11.4	38.9

Had the rate of nuclear energy increased at this rate from 1994 until 2022, nuclear power would have accounted for 1.9 exajoules of energy rather than the 0.8 and accounted for 15.6 % of Canada's primary energy rather than 5.5 %.

Had our adoption of nuclear energy increased as it did in this time period, we could seriously talk about cutting back on the use of fossil fuel. I do not believe there would even need to be a discussion, as nuclear energy generating costs are close to those of hydro power generation. Nuclear energy has some significant advantages over fossil fuel. One of the advantages is transportation. A 1000-MW coal-fired power plant requires a railroad to bring 9,000 metric tons of coal per day to the plant. The daily supply train needs to be 90 rail cars long.

All this coal does not burn. One ton of coal produces about 400 pounds of ash. Ash is toxic, containing a large number of pollutants including lead, mercury, cadmium, and radium. There are airborne emissions of particulate matter as well as gasses. The particulates contain the pollutants already mentioned. The vapours contain mercury that finds its way into local food supply. The ash removed from the coal pants is often put into ponds near the plant. The American EPA has found 90% of these ponds have leaked into the ground water. There was a case in 2008 in Tennessee where the spill of an ash pond required a clean-up crew of 900 men, some developed long-term health problems. There are claims that some workers died from exposure to the chemicals in the ash.

This was why coal was being phased out when there was an alternative source of energy available. Incentives were not required.

Oil and natural gas-burning power stations require an equivalent quantity of fuel, but it is usually delivered by pipeline. There is significantly less by-product as the oil and gas are more refined than coal.

In contrast, a 1000-MW nuclear power plant will require about 30 metric tons of uranium fuel per year. The fuel to power a nuclear plant for six months could be transported to the reactor in a single transport truck. The new fuel bundles for a non-enriched fuel reactor can be transported without special precautions. All Canadian nuclear generators use non-enriched fuel. The waste from a nuclear plant requires special precautions. Nuclear waste will be discussed later in this section.

What happened in 1993 that resulted in a drop in the energy supplied by nuclear energy? The drop actually began earlier. All machinery has a lifetime and eventually has to be taken out of service. The effect of the anti-nuclear sentiment was not felt until nuclear plants had to be taken out of commission and there were no new projects to replace the facilities that were shut down.

The anti-nuclear movement began as the 'Ban the Bomb' protests in the late 50's and early 60's due to the devastating effects of atomic bombs. With the Cuban missile crisis fresh in everyone's minds, the public was shocked to find that above-ground nuclear bomb testing had bathed the globe in radioactive fallout. This only became public knowledge when an American photographic film manufacturer noticed film was becoming partly exposed before it had been exposed to light. The scandal that followed tainted the public's already-bad opinion of atomic bombs, and nuclear energy in general. The public lost confidence in the military and government agencies.

Children were told not to eat snow because it contained fallout. Traces of radioactive isotopes were found in milk.

Not everyone was convinced by assurances from agencies already held in suspicion that nuclear power was safe. In 1979, there was a well-publicized accident at the Three Mile Island nuclear power plant. The incident was fictionally dramatized in a popular movie "The China Syndrome'.

In 1986, at Chernobyl in the USSR there was an explosion. It was essentially a large steam explosion, but it damaged the reactor and a large amount of radioactive material was blown into the surrounding countryside. Workers and clean-up personnel died from the effects of acute radiation poisoning. The explosion blew radioactive material into the air, contaminating the area around the plant. Some material was blasted into the upper atmosphere and fell in countries beyond the borders of the USSR. Chernobyl was a true nuclear disaster.

Both these accidents put a pause on new nuclear projects around the world.

In 2011, an accident at the Fukushima Daiichi nuclear plant in Japan was caused by a natural disaster. An earthquake and subsequent tsunami cut off power to the cooling pumps and the reactor overheated. Following Fukushima, some jurisdictions began to shut down nuclear plants before their lifetime, simply as a precaution. The 911 World Trade Centre attack raised the specter of a terrorist attack against a nuclear power station.

With the exception of Chernobyl, there have been only three deaths due to nuclear accidents in power plants around the world since the early 1960's. There have been some non-nuclear industrial deaths at nuclear power plants caused by accidents such as steam explosions.

The Chernobyl disaster stands out. The accident was the result of a flawed reactor design used only in the Soviet Union. At the time, the Soviets had isolated themselves and did not share technical information with the international community. President Gorbachev was quoted as saying the Chernobyl accident was an important factor in the fall of the Soviet Union. Twenty-eight people died within weeks of the disaster, and as many as 1000 emergency and clean-up personnel received high doses of radiation. Those that died after the accident due to the effects of radiation brought the nuclear death toll up to 62 people.

Grim as these figures are, they would have to be compared to accidents or chronic ailments involving coal and its use.

Nuclear power plant construction has slowed, but a considerable amount of international research has gone into developing more efficient and safer nuclear generators. Canada was a leader in the development of reactors that could be refuelled under load and did not require enriched nuclear fuel that is associated with weapons technology. Canada has an impeccable safety record. More than 50% percent of the electricity produced in Ontario, Canada's largest and most populated province is generated by CANDU type power plants. A research program called The International Gen IV research effort has learned from the history of early reactor designs. Gen IV reactors are designed against natural disasters and even terrorist attacks. The Fukushima reactors are considered to have been based on second generation reactor designs.

Nuclear Waste

A CANDU type nuclear reactor uses natural uranium (unenriched) in fuel bundles. The reactor consists of horizontal tubes that pass through the reactor. Fuel bundles are pushed into the reactor at one end. This is accompanied by the removal of a spent fuel bundle at the other end. The bundle remains in the reactor for from 12 to 18 months. When it comes out, the spent fuel bundle is radioactive. It is taken out by a machine and placed in what is essentially a deep swimming pool. The water cools the spent fuel bundle and the depth of water prevents radiation from escaping. After a cooling-down period, the bundles are moved into canisters and stored underground.

There can be complications with this carefully-controlled process that has to handle about 15 kilograms of material per week. This complexity does not compare with processing 1,500 metric ton of ash taken out of a large coal-fired power plant every day.

6) What is net-zero?

Net-zero means that you can emit as much CO_2 as you always have, but you will have to pay for the privilege. You pay by buying carbon offset credits.

The owner of a wood lot can sell carbon credits equal to the amount of CO_2 the forest absorbs. Let us take a closer look at this. A natural phenomenon takes CO_2 from the air and converts it into plant matter and oxygen. This has taken place for billions of years and will take place for another billion years. It can now be sold. It is like selling air. It is because of the ability to buy and sell CO_2 that the focus of environmentally-conscious people and organizations has been nudged in the direction of considering CO_2 to be the root of all evil. Finding a solution to using fossil fuel would result in a reduction of CO2 emissions. This solution to the problem does not provide an opportunity for speculators.

Some vital industries have to emit CO_2. Cement is one of those industries. Basically, limestone is heated to turn it into lime. CO_2 is given off, and fuel is burned to heat the limestone. Almost 1 metric ton of CO_2 is emitted for every tonne of cement produced. It is the nature of the process. Based on current rates, cap-and-trade will add $35 to every ton of cement produced. The steel industry produces about 1.4 tons of CO_2 for every ton of steel. It is part of the process that converts iron ore into iron metal. It cannot be reduced. These industries cannot be persuaded or coerced. No lawyer has ever been able to strike down a law of nature.

These industries will have to buy carbon credits.

Your first reaction may be that it serves the smelly industries right. Make them pay. But industry will not be left carrying the can. You will. Industry will pass the extra cost on the customer, just as they do with all other costs.

Steel and concrete are essential for building housing, hospitals roads and bridges. Reinforced concrete sewer pipes are used to transfer sewage to sewage treatment plants.

With cap-and-trade in place, you will pay a premium on every twist tie used to package bread, every tin can, every knife, fork, infant chair or car you buy. All these nickels and dimes will roll into the pockets of the cap-and-trade brokers.

In my opinion, this is the perfect crime. Rob the poor to pay the rich with the support of a government in a democracy that should be protecting the interests of Canadians. It is only possible because we are looking at CO_2. This gas is the symptom of burning a non-renewable resource. It is being treated as if it were the disease.

Net-zero does not stop with woodlots.

The owner of a wind power station connected to the grid can sell credits based on the installed capacity. This sale may be possible even if the wind farm is not producing energy that can be sold.

Natural gas pipeline operators have always added hydrogen or inert gasses to natural gas to ensure the energy contained in a cubic foot of natural gas is 1,000 BTU. This is the basis upon which natural gas is sold.

Making hydrogen by electrically separating water requires more energy than can be obtained from the hydrogen. Nonetheless, carbon credits can be sold based on the amount of 'green' hydrogen sold to the natural gas pipeline operators.

Net-zero is big business. In 2022, the carbon offset business was estimated to be worth 332 billion US dollars. It is a growing industry that rides on the shoulders of a population accepting personal guilt for climate change.

What will net-zero do? From an environmental point of view, it will do nothing. A company or industry can still emit, but now they have to pay for it. The natural processes that consume atmospheric CO_2 are functioning as they were before net-zero. This is an opportunity for a small group of people to make fortunes of money. It is a tax opportunity for the government. You and I will pay for price increases on the cost of everything we buy.

Analysis of countries that have adopted cap-and-trade policies has not shown a reduction in emissions. This is not surprising, as cap and trade is a shell game that shuffles money around but provides no alternative to using fossil fuels.

This is like giving someone a $5 per day fine for every day a screw is in a board and promising that person $100 to take it out. It there is no screwdriver available, the person will pay $5 per day and pass the cost on to his customer. The incentive may be there, but the solution is not.

Cap and trade is not a solution. Money pours in, with industry forced to pay. You will never know where it goes. Industry passes the cost on to the customer.

The only reason this shell game works, and is profitable for the few is that it does not address a real problem. The real problem is that we are running out of fossil fuel and we need to find an alternate source of energy. Cap-and-trade forces us to focus on the CO_2 which is in reality an artificial problem that allows the cap-and-trade industry to exist.

There is a lot of money at stake in the cap-and-trade industry. All of it is dependent upon the public perception, and fear that CO_2 has to be stopped. Net-zero is based on our belief that we all have to do our little bit to stop climate change. There is no room to muddy the waters with solar angles, solar storms, water vapour, methane or geological history.

You may read this and ask what I know? After all, international conferences attended by delegates from almost every country on earth see net-zero as a way to lower CO2 levels in the atmosphere which will save the world.

My opinions do have some international support. During the COP 28 climate conference in Dubai the conference president, Sultan al-Jaber was criticized for his statement that there was no science behind calls for a phase out of fossil fuels. He was criticized, but no one stood up and pointed to the science he claimed did not exist.

I am making the same statement. I maintain there is no sound science supporting the relation between CO_2 and climate change, however there is a practical reason why the world must reduce fossil fuel use. Based on estimated world reserves of oil, coal and natural gas we have from 60 to 100 years of fossil fuel left before we run out. This is based on today's rate of consumption which has increased every year since the 60's.

That is why we have to find an alternate to the use of fossil fuel.

Fretting about CO_2 and taxing its production and considering net-zero as a solution would be like a passenger on the Titanic telling the maître d' an ice berg was not a good reason to delay dinner.

Forget CO_2. We need to develop a strategy to replace fossil fuel. We could start with the portion used to generate electricity. Given the advertising pictures of electric cars driving past wind farms, it is hard to believe that 61% of the world's electricity is generated from fossil fuel. During 2022 this broke down to 35% from coal, 23% from natural gas and 3% from oil. Converting to nuclear power generation would stretch out our reserves to allow more time for the development of portable energy options.

It behooves the government to examine the concept closely and determine if net-zero is a program that is in the best interests of Canadians based on facts that can be supported.

7) The cost of a throwaway society

Adopting Nuclear energy as a major source of global energy is a project that will take 20 to 30 years. Financing has to be arranged. Sadly, it is more time consuming to finance a public works project than a war. Putting together financing for a major project such as a 2000 MW nuclear generation station could take years. Planning and design take another two years and construction may take an additional 2 to 3 years. In practical terms, the global community can only build a limited number of major projects at a time due limitations such as engineering capabilities and construction skills.

While we go about this process, can we reduce energy consumption? Of course we can, but beating up or even subsidizing consumers is not the way to do it. Certainly, energy can be saved with improved insulation in homes. Energy can be saved by installing heat pumps.

Government programs have already addressed these ideas, but the largest potential for reducing energy consumption can be achieved by reducing the need to manufacture. We live in a throwaway society, not through consumer choice by through greed. Things must break in order to force consumers to buy more. This allows corporations to show growth.

A good example are light bulbs. If you grew up with incandescent or glowing filament light bulbs, you were accustomed to lights burning out. There is a technical reason why filament type light bulbs had a limited life. Old-style light bulbs used only a small fraction of the energy they consumed to produce light. The rest was lost as heat.

LED (Light Emitting Diode) light bulbs are far more efficient. It takes 6 to 8 watts to produce as much light as a filament light that drew 60 watts. LED's have a far greater advantage. They do not burn out. Most people have some form of electronic device over 10 years old with its original LED happily glowing as it was the day you first plugged it in.

Manufacturers are permitted to sell LED lights designed to burn out. LED light bulbs fail because the circuit is designed to draw more power than the devices are rated to handle. This is why they get hot. Not only does the generation of heat waste energy, but the heat eventually causes the LED or a component in the relatively complex circuit in each bulb to burn out.

When the first LED bulbs came out there was a 16-year guarantee written on the package. You do not see that anymore; however, it could be made a condition of selling LED bulbs in Canada.

We have no consumer regulations to protect the consumer from intentionally wasteful designs. We have carbon tax.

Generally, solid state electronics do not fail. The NASA satellite Voyageur I was launched in 1977 and is now at the edge of the solar system. It stopped transmitting data in December 2023 after 46 years. Electronics that are flying past Jupiter are in a more hostile environment than components sitting in your living room or in your car. All Solid State electronics could be designed to last until you get tired of using them. That includes electronics in cars.

There is no reason why a refrigerator, washer or dryer could not be a once-in-a-lifetime purchase. Most major appliances fail because the electronic controller stops working and cannot be replaced, or the replacement is the same price as a new machine. There is no reason for this. There is no reason why you should not be able to purchase a new controller at a reasonable price. There is no reason why small mechanical components should not have a long life, after which replacements should be possible.

Cost effective repair and replacement, even upgrades are possible. An example is electronic ignition. The points and capacitor system used in older lawn mowers was replaced by electronic ignition. This is a more reliable ignition system. All new lawn motors have them. One manufacturer was able to make their electronic ignition module the same size as the points and capacitor system. For a very low cost an electronic ignition module can be fitted to an old lawnmower.

Automobile manufacturers have more contempt for the consumer than the makers of lawnmowers. Today it is not possible to re-use an alternator or starting motor in a newer car, even if the old car is the same make and model.

All an alternator does is generate 12 volts to charge a battery. All a starting motor has to do is generate enough torque to turn over a car engine. Not only could they be interchangeable, but it would not be difficult to have a standard design that would fit in all cars regardless of the make. If this were the case, a replacement starter or alternator would not cost in excess of $350, it would cost about $50; the cost of materials and manufacturing with a reasonable profit for the manufacturer.

From the 1940's almost until the 80's all cars from all manufacturers used the same size of headlights. These were interchangeable, and they contained the front lenses. It cost about $5 to buy a new headlight which a car owner could replace with a screwdriver. Now headlights are moulded plastic that, after time, clouds over and must be replaced at a cost of several hundred dollars. Standardization is possible, but the car manufacturers can get more of your money if everything is different, and we have no agencies or regulations forcing standardization.

Electric cars are worse. The entire car is a throwaway. The battery replacement can cost almost as much as the entire automobile. We are beginning to learn that a battery – hence a car replacement is necessary if the car is involved in a fender bender due to the risk of spontaneous combustion of the battery. There are not many things an independent garage can do with an electric car because the manufacturers will not release repair software or testers. Regular cars are following this trend and there are no regulations protecting the consumer.

All this has an energy cost. A component of the increasing global energy demand is driven by our throwaway society, which is driven by greed. Not your greed, but the greed of a vanishingly small number of people at the top of large corporations.

How much does this cost?

Energy is in reality a form of international currency and I will talk about kWh and MJ rather than dollars.

Studies are becoming more common in which researchers examine products and look at the energy requirements from the mining of raw materials through the eventual scrapping of the product.

As an example, researchers studied an economical, gasoline-powered car in 2011, weighing 1,480 kg. They traced all components back to the source and found the 1,480 kg automobile required 7,762 kg of raw materials. This is due to, for example, the amount of iron ore that must be extracted to produce the steel in the car. The authors of the study found that it required 62 Gigajoules to produce a 2011 gas-powered economy car. This is 4.81 MJ per kg of car weight.

Appliances like refrigerators and stoves have a similar high energy cost of production and shipping, often from across the globe of about 2,000 kWh per fridge or stove. This is not far from the energy a household uses in a month. If a refrigerator lasted 15 years rather than 10, there would be an annual energy savings in Canada of approximately 400GWh. This would be felt not only in Canada, but it would reduce global energy consumption by this amount.

The effect of increasing the life of an automobile would be dramatic. Considering fabrication energy for small gas-powered cars, if cars in Canada alone could last 15 years rather than 10, the global energy saving would be in the order of 4.5 TWh per year. Once again, this would be a cash benefit to Canadians and a meaningful reduction in global energy consumption—a win-win situation for the environment and Canadians.

  Electronic component manufacture requires a tremendous

amount of energy. It requires about 2 GJ to make a 300-mm wafer that is the basis of all electronic components. A single processor chip requires 3 MJ. The manufacturer of a cell phone or a small computer could consume as much as 5MJ, or 1.4 kWh If cell phones and computers in Canada alone lasted 10 years rather than 3, the savings to global energy consumption would be 6.5 GW annually.

This would be possible with consumer protection laws that supported the right to repair.

Small businesses should be able to repair cell phones. As I write you cannot even swap a component from one identical phone to another. The phone will recognize a new component and prevent the owner from using the phone. You cannot swap a good component out of a cell phone to replace a faulty part. Factory service is almost impossible to access, so people have to buy new, usually more expensive phones whether they want a new phone or not.

Fifteen years for a car and major appliances and ten years for cell phones are not unrealistic goals. The cost of slight improvements in components such as bearings and onboard electronic reliability would be measured in dollars per car or appliance. Corporate growth that only kept pace with the growing population would have an impact on a limited number of top executive's annual bonuses.

An important secondary benefit would be the use of fewer raw materials. This would allow raw material costs to drop reducing the price of the product.

By protecting Canadians with reasonable product lifetime requirements, and reasonable repair part availability Canada could have an impact. Right to repair legislation would need to be improved, and software compatibility would be needed. Both these initiatives would benefit the public, reduce costs and would not have to be designed to make people feel guilty enough to pay carbon tax or help buy fuel for a cap-and-trade billionaire's yacht.

8) Where do we go from here?

The problem our society faces is not CO_2, or climate change. The real problem we face is that there are 8 billion of us on earth and we are entirely dependent on a fuel source that is running out. 81% of the world's energy comes from fossil fuel, and that is not declining. I believe that the solution will be to begin building nuclear power plants so we can reduce our dependence on fossil fuel.

We have no choice. We cannot go back to living off the land. Population densities for hunter-gatherers were 21.6 per square mile. 8 billion people cannot be spread that thin.

Can we shut down all the dirty smelly factories? Not unless we want to return to the stone ages, which is not an option. There are already too many of us. Factories exist to make things we need to live. We cannot build hospitals or even housing without steel and cement. We cannot store and transport food without refrigeration. We need to build and operate ships, trucks or other vehicles for transport. We cannot farm to feed the world's population without tractors and farm equipment to produce the food.

We need energy to heat our homes, offices and factories. Without mechanical cooling, some of our major cities such as Atlanta, Hong Kong and even Washington would be almost uninhabitable during part of the year.

Our society cannot survive without available, inexpensive energy. Without a continued supply of energy our society will collapse. If we run out of energy, we will not see a gradual slow down. There will be increasing world conflicts and increasing refugee crises. You might not be wrong to say this has already started.

The solution does not require a lot of new money, just some refocusing.

1) We have to move away from the concept of net-zero. It may be argued that this was developed by international co-operation at international conferences. I do not believe these

initiatives are in the best interests of Canada and Canadians. Some European countries are re-examining the benefits and their commitment to the scheme.

2) Develop a road map toward the promotion of new nuclear power generation in Canada. We are well on our way, and have historically been a world leader in the peacetime use of nuclear energy. We need to examine the process and ensure there are no barriers to building nuclear power plants, while at the same time maintaining the same public safety standards already in place.

3) Strengthen consumer protection laws to include minimum standards for product life. Industry standardization would open up a market for small to medium size companies to make aftermarket replacement parts. Right to repair legislation needs to be strengthened for mechanical equipment as well as for electronics and software.

4) Drop carbon tax. It clearly has no measurable effect otherwise the benefits would have been trumpeted loudly when the credit checks were rolled out. Carbon tax raises the price of everything we buy. Years from now an auditor will no doubt announce that Canadians got back penny's on the dollar. Governments around the world are notoriously inefficient, in spite of everyone's best efforts. The best place for your money is in your pocket.

If all these suggestions come to pass, will we be able to eliminate the use of fossil fuel? I do not think so, but its use can be reduced to the point that world reserves will last for many generations until substitutes can be developed for the special cases. Will CO_2 emissions continue? I believe they will be reduced, but not eliminated.

With this reduction in fossil fuel CO_2 production, will the climate continue to change? Without a doubt. History has shown that the climate has always been changing. These changes will continue to occur with or without human intervention. CO_2 is a red herring.

Keep in mind: climate change is big business. In order to get your money, you are being convinced you are responsible. Current climate initiatives such as cap-and-trade provide no more solutions than a tithe. Initiatives that focus on CO_2 are designed to take money out of your pocket and lower your standard of living. These initiatives have been designed by special interest groups to further their own interests. If we do not turn our attention to developing our nuclear power industry, our children will have no future other than the life of the dark ages: a life of manual labour and continuous war.

Appendix 1

Multipliers

Prefix	Symbol	Scientific equivalent	Spread sheet representation	multiplier
kilo	k	1×10^3	1.00E+03	1,000
mega	M	1×10^6	1.00E+06	1,000,000
giga	G	1×10^9	1.00E+09	1,000,000,000
tera	T	1×10^{12}	1.00E+12	1,000,000,000,000
peta	P	1×10^{15}	1.00E+15	1,000,000,000,000,000
exa	E	1×10^{18}	1.00E+18	1,000,000,000,000,000,000

Conversion factors for energy

from	to	Multiply by	spreadsheet
joules	BTU	9.49×10^{-4}	9.49E-04
BTU	joules	1.06×10^3	1.06E+03
joule	KWh	2.78×10^{-7}	2.78E-07

Calculated CO_2 generation from fuel table 1-1

example

World energy from coal 2022:   161.47 Exajoules – source: Statistical review of world energy

https://www.energyinst.org/statistical-review

Conversion factor exajoules to million BTU multiply by 9.48×10^8

161.47 EJ X 9.48×10^8 MBTU/EJ = 1.53×10^{11} million BTU

Coal (all types) 95.92 Kg CO_2 per million BTU Source: https://www.eia.gov/environment/emissions/co2_vol_mass.php

1.53×10^{11} million BTU X 95.92 Kg CO_2 per million BTU = 1.47×10^{13} Kg CO_2

1 metric ton (can be written as tonne) = 1000 kg

1.47×10^{13} / 1000 = 1.47×10^{10} metric ton CO_2

1.47×10^{10} / 1,000,000 = 1.47×10^{4} = 14,680 million metric ton of CO_2 from world coal consumption in 2022. This number appears in table 1-1

Appendix 2

CO_2 from forest fires during 2023

Living tree is 50% water and 50% carbon. Source: Illinois dept. of natural resources

An average of 90 Tons of timber can be harvested per acre when clear cutting.

Add 75% extra for leaves and branches = 67.5 tons leaves and branches.

90 + 67.5 = 157.5 tons of wet wood per acre.

Dry weight = 1/2 wet weight = 78.8 tons of dry wood

This is 50% carbon = 39.4 tons carbon per acre.

$C + O_2 = CO_2$

Atomic Weight C = 12, Atomic weight O = 16

Molecular weight CO_2 = 44

12 tons C = 44 tons CO_2, so 1 ton C produces 3.67 tons CO_2.

Carbon, something you can hold in your hand, when burned produces more than its own weight of a gas that under normal circumstances you cannot weigh.

39.4 ton carbon per acre produces 39.4 X 3.67 = 144.6 ton CO_2 per acre if the wood is burnt to ash.

In 2025, 14,000,000 hectares of forest were consumed by wildfires.

1 hectare = 2.47 acres so 14 million hectares = 34.5 million acres

34.5 X 144.5 tons CO_2 per acre = 4.996×10^9 tons CO2

* a ton is 2,000 pounds, a tonne or metric ton is 1000 kg =

2200 pounds.

4.996×109 tons X 2000 /2.2 = 4.5×10^{12} kg CO_2

4.5×10^{12}/ 1000= 4.5×10^9 metric tons

 4.5×10^9 /1,000,000 = 4.5×10^3 million metric tons = 4,500 million metric tons CO2 from burning 14 million hectares of forest.

Appendix 3

CO_2 from human respiration

Standard respiration frequencies and volumes.

https://www.ncbi.nlm.nih.gov/pmc/articles/PMC8672270/				
Condition	Rate (Breaths/min)	Tidal Volume (L/breath)	Minute Ventilation (L/min)	l/min CO₂ @ 5%
Nominal At-Rest	12	0.5	6	0.3
Normal Activity	16	1	16	0.8
Moderate Exercise	20	2	40	2.0

https://byjus.com/biology/composition-gases-breathe/

We breath in air containing .04 % CO_2, the air you exhale contains 5% CO_2.

Assumptions about activity used in the calculations

			hours per day	
	Litres/minute CO₂		Assumed hours per day	Volume of CO₂ in litres
rest	0.3	l/min CO2	6	108
normal	0.8	l/min CO2	10	480
active	2	l/min CO2	8	960
		l/day/person		1548

Example .3 l/min CO_2 X 6 hours X 60 minutes = 108 litres of CO_2 produced during the 6 hour period

Convert litres of CO_2 into grams of CO_2 using the ideal gas formula

$PV = nRT$

P = 1 atm
V = 1548 liters of CO2 per day per person
R = .082   l.atm/K.mol
T = 289 degrees absolute (K)
n = unknown

Solving for n = 63 moles per day CO_2

63 X MW CO_2 (44) = 2772 grams CO_2 per day per person = 2.8 kg

Population example:

Canada 39 million

$39 \times 10^6 \times 2.772 = 1.08 \times 10^{11}$ kg CO_2 per day = 1.08×10^5 metric tons CO_2 per day.

$1.08 \times 10^5 \times 365 = 3.93 \times 10^7$ metric tons per year = 39.3 million metric tons of CO_2 produced every year by Canadians breathing.

Appendix 4

<u>Calculation of installed capacity utilization – table 4-1</u>

Example Wind power utilization

There are a number of reliable sources for the data used in this paper. Some examples are presented below. Each reference often provides a slightly different number, however the differences reported are not significant enough to change the conclusions reached. The differences are not large enough to cause the author to question the validity of any data source investigated.

Energy produced 37.15 TWh – reference Statistical Review of World Energy

Energy produced 36.06 TWh – reference NRCan

Installed capacity 15.4 GW - reference Canada wind power market research report

Installed Capacity 16.1 GW- reference Statistics Canada

Installed capacity 15.3 GW - reference www.statista.com

Installed capacity 15.31 GW - reference NRCan

From table 4-1

Installed capacity is 15.4 GW. This is the amount of power all wind turbines in Canada could produce if they were all running at 100% capacity. If they ran 24 hours a day for 365 days, the energy produced would be:

15.4 X 10^9 watts X 24 hours X 356 days = 1.35 X 10^{14} watt-hours.

1.35 X 10^{14} / 10^{12} = 1.35 X 10^2 TWh = 135 TWh

135 TWh is the amount of energy that would have been produced if all turbines could have run at full capacity for an entire year

Energy actually generated was 37.15 TWh

(37.15 X 100)/ 135 = 27.5% utilization